The Two Faced Moon

Simon Lewis

Order this book online at www.trafford.com/07-0480
or email orders@trafford.com

Most Trafford titles are also available at major online book retailers.

© Copyright 2007 Simon Lewis
All rights reserved. No part of this publication may be reproduced, stored in a retrieval system, or transmitted, in any form or by any means, electronic, mechanical, photocopying, recording, or otherwise, without the written prior permission of the author.

Note for Librarians: A cataloguing record for this book is available from Library and Archives Canada at www.collectionscanada.ca/amicus/index-e.html

ISBN: 978-1-4251-2076-4

We at Trafford believe that it is the responsibility of us all, as both individuals and corporations, to make choices that are environmentally and socially sound. You, in turn, are supporting this responsible conduct each time you purchase a Trafford book, or make use of our publishing services. To find out how you are helping, please visit www.trafford.com/responsiblepublishing.html

Our mission is to efficiently provide the world's finest, most comprehensive book publishing service, enabling every author to experience success. To find out how to publish your book, your way, and have it available worldwide, visit us online at www.trafford.com/10510

 www.trafford.com

North America & international
toll-free: 1 888 232 4444 (USA & Canada)
phone: 250 383 6864 ♦ fax: 250 383 6804 ♦ email: info@trafford.com

The United Kingdom & Europe
phone: +44 (0)1865 487 395   local rate: 0845 230 9601
facsimile: +44 (0)1865 481 507   mail: info.uk@trafford.com

10 9 8 7 6 5

Contents

Foreword	5
Introduction [The Two Faced Moon]	7
Chapter 1. Alive and Kicking?	9
Chapter 2. The Big Secret	25
Chapter 3. Life On The Moon	43
Chapter 4. The Conclusion and Weird Words	53
Points to Remember	59
The Apollo Astronauts	61
Acknowledgments	63
Pictures	65
The Two Faced Moon [Short movie script by Simon and Carol Lewis]	81

Foreword

Well where can I start? This subject is one of many I became interested in and always wanted everybody else to understand.

I started off as a UFO [Unidentified Flying Object] researchers and joined a group [NORTH LANCASHIRE UFO GROUP]. In this group I met some fantastic people, who all had one object, "to reveal the truth". That truth was that we were living in a world where some people liked to keep big secrets from the public. The subject of UFOs was without a doubt one big cover up from the start and left you always searching for more information.

People out there were seeing things that defied belief and were left with lives that were changed forever. The governments as always denied everything and always washed their hands clean of a situation. You were always left, without a doubt, that whatever was going on out there in our air space, the governments knew about it.

Being involved in subjects such as UFOs soon connected you with people who had either been involved in secret projects or knew exactly where to get the information from. Putting aside Black Projects [Secret Military Projects] it became rather difficult not to think that we were alone in the universe. Many sightings and even footage showed manoeuvres in the air, which defied gravity and would have left most human pilots in pieces and when I say pieces I am referring to body parts. Moving on from UFOs it became apparent through many sources and specialist books that our Moon was not what it seemed.

One day I came across two books, written a long time ago, one was by George Leonard and the other by Fred Steckling. These two men had written books about a Moon that was very much full of activity on the surface. They had boldly suggested at the time that our Moon was inhabited. Well if this was true, then why all the secrecy? And that's where my journey began. It did not take long before I had contacts and had even come across unusual film. What I had found and was told had me hooked. The Moon was and is inhabited.

All I can say to everybody who reads this book is to remember that you only see one side of the Moon. You can believe what you want, but please think carefully, as we all might have been taken for a ride, over the truth of our little neighbour.

The Two Faced Moon
Introduction

When I considered writing this story I was not sure whether it was a good idea or not as the subject would obviously create debate and maybe open a few door ways to the truth. That truth of course could in itself cause great upset to those who would rather stay silent. What if the Apollo Moon landings found something that was so amazing that a cover up would be necessary? Because what had been found would certainly change history.

What if the Moon was not the supposed dead body we make it out to be? And what if the Moon was not really ours?

I intend to open your minds to the facts that we found evidence of life, past or present and reveal startling evidence to back up everything I say.

Welcome to "THE TWO FACED MOON".

Chapter 1
Alive and Kicking?

ON THE 20TH JULY 1969 THE whole world watched live on television as Neil Armstrong set foot on the Moon. Armstrong was an ex-aviator veteran, of the Korean war and former pilot of the Gemini 8 missions. His skills and experience finally gave him the position of commander of Apollo 11, where he was joined by crewmates Micheal Collins and Buzz Aldrin. Armstrong and Aldrin landed on the Moon in the Lunar Module [Eagle] whilst Collins orbited the Moon in the Module [Columbia]. Armstrong will always be remembered for those famous words, "That's one small step for a man, one giant leap for mankind ". The rest is history, or the history they wanted you to hear.

Long before any Moon landings of any kind, observations of unusual activity had been recorded on the Moon. From this came the term, "Chronological reported lunar events", the Moon it seemed was not the dead body we were all made to believe. The official report was documented by Patrick Moore and Barbara Middlehurst in 1968 and updated in 1971. It is an amazing insight into Moon activities over the centuries. As early as 1671 on the 12th November, D Casino observed around the crater P Tatus, small whitish clouds, and in more recent years the NASA Lunar orbitor [Plate 48 NASA Lunar Orbitor V NO MR 168] photographed what seemed to be cone shaped clouds as well as Cirrocumulus Clouds in an area of the crater Vitello, Mare Imbrium. The crater has a diameter of 30 miles and walls rising 4,500feet and is surrounded by small hills and craters. It would seem almost ridiculous to suggest we were observing a warm frontal system but the clouds do seem to show a rippling effect as they pass over the crater. NASA refers to the picture as a picture with marks on it. What the picture does create is argument and big debate because those marks could possibly be clouds. This picture is one of many unusual pictures showing strange events on and around the Moon and this makes it harder to explain them as just being errors or marks or in some cases transmission problems. The crater Alphonsus has had well documented evidence of unusual blurring when monitored. N.A Kozyrev, of the Crimean Astrophysical Observatory in the U.S.S.R. was monitoring the crater Alphonsus using a 50inch reflector on the 3rd November 1958, when he noticed that the central peak had become blurred. What had been quite notable was the reddish cloud on the central peak which soon went very bright and then dim. Alphonsus soon returned to its normal self.

It seems observations by astronomers in past centuries were well documented and a good record kept. One of the main talking points is red –spots, red streaks, bright areas etc. These have been observed around craters such as Plato and Aristachus.These observations go back hundreds of years and these activities lead us to ask detailed questions.

There have been more than 300 reports relating to the Aristachus area alone. Aristachus stands on a rocky plateau, with walls 2000ft high. As early as March 1783 W. Hershel and Lind observed red spots within the vicinity of the crater. The observations continued, Shroter and Van Bruhl from 1784 to 1787 noted unusual bright areas of light in and around craters. Then on 13th Febuary 1835, again unusual bright spots were observed by Gruithuisen. It is almost tempting to say going to the Moon in 1969 was more than just a Cold War statement but a visit to investigate the observations that had been showing, a Moon with many secrets to be told or not to be told. On and on the sightings continued through the decades, all being recorded. In 1963 on the 30th October two observers, Greenacre and Bom, saw three-red spots, one inside Aristachus and the other two nearby in the Schroter Valley. The entire display lasted over 45 minutes.

The Apollo 11 mission even reported a strange fluorescence towards Aristachus, as the following conversation revealed:-

ARMSTRONG: "Hey, Houston, I'm looking North up towards Aristachus now and I can't really tell at that distance whether I am looking at Aristachus, but there's an area there that is considerably more illuminated than the surrounding area. It just seems to have a slight amount of fluorescence to it."

HOUSTON [McCondles]: "Roger, eleven, we copy."

ALDRIN: "Looking at the same area now, well at least there is one wall of the crater that seems to be more illuminated than the others. I'm not sure that I am really identifying any phosphorescence, but that definitely is lighter than anything else in the neighborhood."

HOUSTON[McCondles]: "Can you discern any difference in the color of the illumination and is that an inner or outer wall from the crater? Over"

ALDRIN: "I judge an inner wall in the crater."

COLLINS: "No there doesn't appear to be any colour involved in it Bruce."

Twenty minutes passed and the signal of the Apollo 11 was lost as it moved around the dark side. When it reappeared, nothing else was said. That particular event you would have thought, would have made great news considering how many observations had been recorded in and around Aristachus, but alas no. You immediately wonder why go to the Moon see amazing things and generally don't make it widely known. My goodness, the Moon is not exactly just down the road, yet the conversation seems to be lost on the general public for whatever reason.

So many areas of the Moon which we can see have had unusual activity, from Prolemoreus, Mare Crisium, Posidonius, Plato-to-Bessel, Tycho, Longrenus and on and on I could go. What is most frustrating to say the least, is that all this has been taking place over such a long period of time, and probably still continues today, and yet the public still do not get these facts. Surely we should be made more aware of what I consider to be very important pieces of information. This information I found hard to get and really not available as it should be.

Undoubtedly NASA and the Russians, photographed and most probably filmed unusual phenomena on the Moon, which has never been made public knowledge, and seems to still stay a secret rather than to be made an historic scientific find. You feel as though somebody did not want to demonstrate success and that it was best to stay silent rather than create a storm. As you can see from the next conversation the questions outway the answers. The mission was that of Apollo 16, 22nd April 1972:

DUKE: "Tony, what is the other peak?"

CAPCON: "Right of the cosmic ray experiment."

DUKE: "Ok, Ill cross F-11 250 at 15"

YOUNG: "Ok, Houston, I just picked up this white rock, but it has a black layer on the back of it, a thick black glass and its about[garble] specimen."

CAPCON: Scramble. "Hey, fellow, Ken was. Just flying over and he saw a flash on the side of Descortes-he probably got a glint of you? "

DUKE: "Oh sure, that's us. Men of miracles. We're dusty."

YOUNG: "Don't step right here, Charlie, there's a splatter, a glass splatter. A whole big bubble of it, isn't it?"

This particular conversation makes it quite clear that the flash seen was not anything to do with the astronauts on the Moon as I am sure they were well trained in observation as well as conclusions.

Back in 1972 NASA announced that it would be studying LTP [Lunar Transient Phenomena]. Basically NASA would require anyone who had the equipment and knowledge, to carry out a study of the Moon in detail. There were thirty two responses. These people would have to study four areas of the Moon that had shown LTP. The sad factor from this gathering was that not enough people reported back so it came to a rather abrupt end, leaving us with only the 1968 report and 1971 updates as documented evidence.

The Apollo missions did find all sorts of interesting things which were made public, these were of course the many samples, which included rocks, dust, crystals etc. Thousands of samples were brought back from the Moon, a staggering 75000 pieces. Studies carried out on a majority of the pieces found materials that were composed of glass drops, from a meteorite to a large breccia encased in molten rock. Many of the materials would be ideal for industry, silicon for computers, soil which could be used to make glass and all in a low gravity enviroment. What we must look at are the abundance of materials where energy comes into the equation.

Helium 3, which comes from Moon dust, recently found by NASA scientists holds the key. Dr John Santarius of Wincinson University has come to the conclusion that 25 tons of Helium 3 could power the whole of the United States for a year. When you consider the Space shuttle payload bay can carry that amount things become more puzzling or coincidental.

It is rather amazing that the Moon is more of a secret now, than it was before we landed on it. There are issues here of course, perhaps our close neighbor has more to offer than we would like to make public. The information that is released I always find annoying, it's always small news, never a headline. In some cases it does make the front pages of the newspapers, as with the case of water being found on the Moon. What is interesting is how a number of astronomers and scientists in the past clearly said we would never find water on the Moon. On the 3rd December 1996 frozen water had been found deep in a crater on the south pole. The probe [Clementine] which was military I must emphasize, discovered an area of water twice the size of Puerto Rico with a depth of 1.3 kilometres. This finding of water changes everything if you want to colonise the Moon. This whole thing about water reminds you that it was actually discovered back in the 70s on the Apollo missions and caused great argument at the time. NASA Assistant Director of Lunar Science Richard Allenby quoted," There is no evidence in the rocks or geochemistry that water exists". This statement soon was to be squashed when Dr. John Freeman, and Dr. H. Ken Hills announced that great eruptions of water vapor clouds had been detected which covered an area of 100 square miles of the Moon's surface and had lasted 14 hours. When you examine the strange photographs of clouds seen on the Moon you can easily make the connections. NASA decided to explain the whole event as being water tanks off the Apollo descent stages. The science team reacted by pointing out that both Apollo 12 and 14 had been located over 180km away so they could not have brought the water themselves. Things really started going downhill when NASA announced it was merely pee released from the space capsules. The Apollo 17 mission on December 29th 1972 reveals once again the whole argument over water.

CAPCOM: Roger. America, we're tracking you on the map here, watching it.

LMP: O.K., Al Buruni has got variations on its floor. Variations in the lights and in its albedo. It almost looks like a pattern as if the water were flowing up on a beach. Not in great areas, but in small areas around the southern side, and the part that looks like the water-washing pattern is a much lighter albedo, although I cannot see any real source for it. The texture, however, looks the same.

CAPCOM: America, Houston. We'd like you to hold off switching to OMNI Charlie until we can cue you on that.

CDR: Wilco.

Amazing stuff, but as we get back to the conversation the word water is mentioned again.

LMP: O.K 96:03. Now we're getting some clear –looks like pretty clear high watermarks on this—

CMP: There's high watermarks all over the place there.

LMP: On the north part of Tranquillitatis. That's Maraldi there, isn't it? Are you sure we're 13 miles up?

CAPCOM: You're 14 to be exact, Ron.

LMP: I tell you there's some mare, ride or scarps that are very, very sinuous- just passing one. They not only cross the low planar areas but go right up the side of a crater in one place and a hill in another. It looks very much like a constructional as I would want to see it.

So we have a not-so-dead Moon as most of us seem to think, in fact we have a Moon that is very much alive or am I just trying to bend the truth and make you all go down the wrong avenue, I don't think so! Things just do not make sense, so many astronomers and scientist seem to want to stand away from the crowd and yet something holds most of them back. The Moon is so close and yet so far, it holds more secrets but has the answers we are looking for. The whole water incident then was just brushed aside. Decades later we find water on the moon officially accepted, and found by Clementine.

Without a doubt technology has brought the Moon closer than ever with the advance of computers and general telescopes. The telescopes available to the general public are superb and can see most night sky objects with a trained eye and patience. Computers and digital technology combined to give us pictures and film. One of the many questions that pass through my mind is how your average person with this new machinery does not seem to report much. With the Moon being so close you sort of wonder if the Moon is just as most scientists say, dead! But remember it does have a hidden side and just as the space shuttle has shown when in space looking at Earth there does not seem to be life, at first glance, on our planet, it prompts me to think its time to go back to the Apollo missions and look at what was really going on. In the next chapter the questions outweigh the answers far too much, while the conclusions speak for themselves.

Earth Clouds

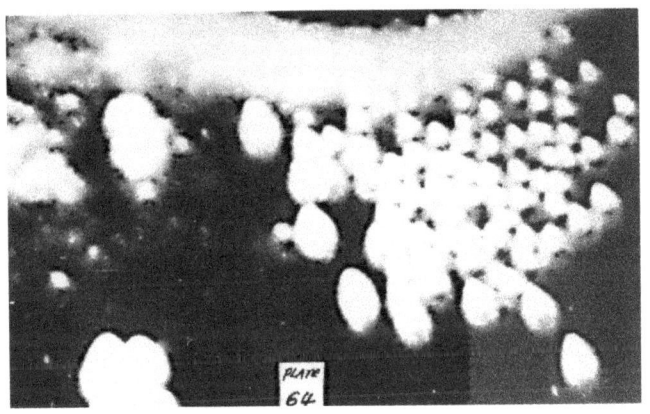

Crater Ritter (area blow-up) showing cloud.
Lunar orbiter photo MR81

Clouds moving over crater vitello.
(NASA Lunar orbiter V NO. MR168)

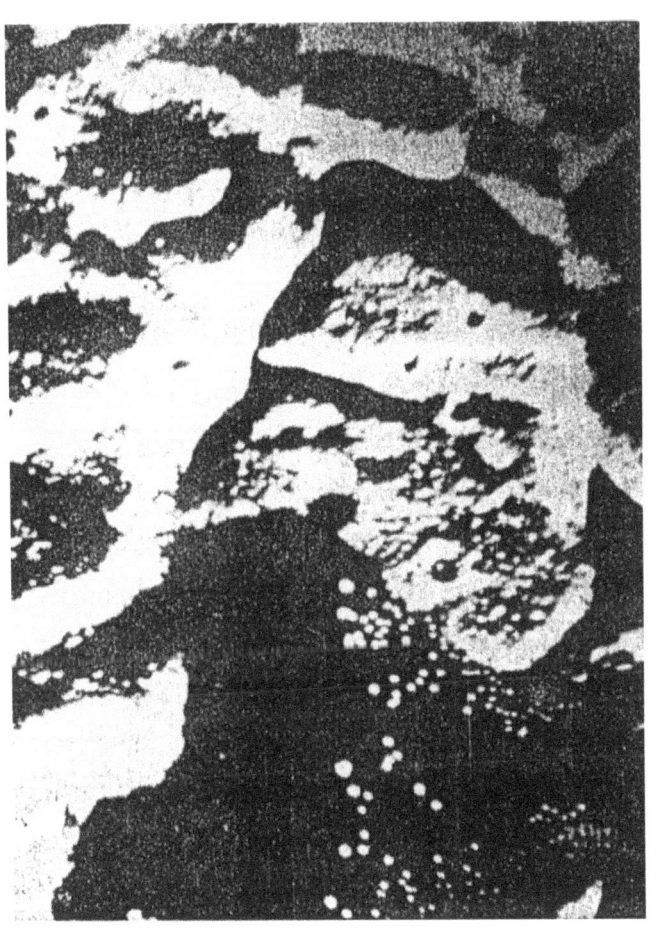

Earth clouds? No moon clouds!
(Area blow up) Vitello to the west.
(NASA 10 V NO MR682)

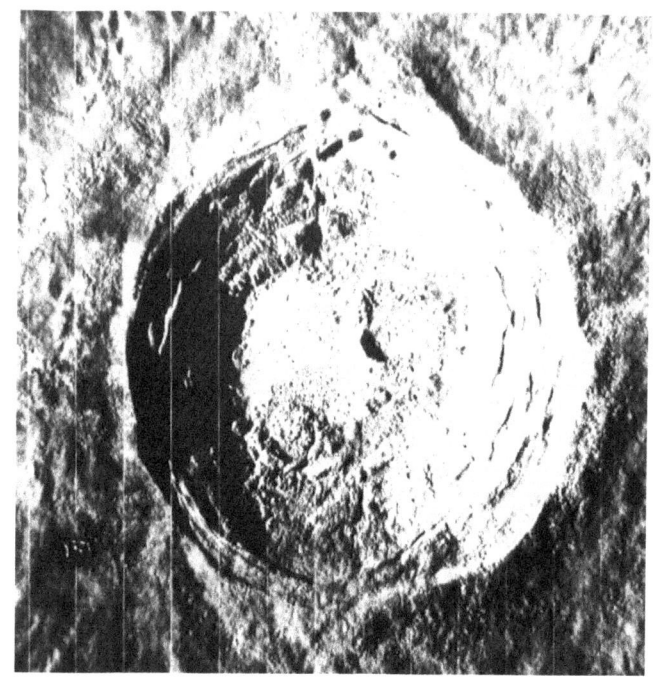

Aristachus a crater of many unusual sightings.

Chapter 2
The Big Secret

SO WHY DID WE GO TO the Moon? Was it a Cold War thing or maybe something else? Lets imagine that going to the Moon was more about who got the treasure first. We have already looked at the facts on Chronological reported lunar events, which on their own were one good reason to go to the Moon. But what if something else was there, and was not natural and had left its mark and just maybe was still there. Let's go back to one very interesting newspaper article which at the time, had a lot of familiar names involved. The newspaper was the Pulse and on December the 14th 1968 had a lot to say about the landings on the Moon, their headline read "MOON SECRETS SOON TO BE EXPLAINED, Human Landings May be challenged". Sounds rather a bold suggestion but the article continued with a Major Patrick Powers [Head of the United States Army Space Development Program], stating that, "the first man to reach the Moon must be prepared to fight for the privilege of landing". The article continues with Dr Carl Sagan, who back in 1962 was the adviser on extraterrestrial life to the US forces stating that mankind must face the probability that beings from elsewhere in the Universe have, or have had, bases on the averted side of the Moon. Very powerful words for such people at the time who were very much aware of something which was not public knowledge then or now.

I think the most interesting information in the Pulse article is about the artificial bridge. This part of the article reads as follows, "It was the late John O'Neil, former science editor of the "New York Herald Tribune" who really caused the biggest stir in Lunar observations when, in July 1953, he reported what he called a gigantic artificial arch 12 miles long-the Lunar bridge." This was later confirmed by Dr H.P.Wilkins at the Mount Wilson Observatory, USA. A month later, when viewed under good conditions, the bridge seemed to have disappeared. Its so interesting that so many people were hungry for something that the Moon had but until we landed were not willing to share, while at the same time the thoughts that must of being going through people's minds must have been mind blowing.

Finally in the article is the story of the spikes. The article reads as follows,"On November 18th, 1966, a Lunar orbitor probe left Cape Kennedy. Three days later, from a height of 30 miles, the instrument's telephoto lens was focused on the Moons surface and the camera activated. On a small section of the Sea of Tranquility, a plain just off the Moon's centre, the camera picked up six spike-like projections, casting shadows across the dusty surface. They were called, "some of the most unusual features of the Moon ever photographed " by the scientists in charge of the project. However, they felt the spikes to be natural. Mr William Blair, a Seatle anthropologist and a member of the Boeing Company's biotechnology unit, thought otherwise. They were he maintained, in a geometric pattern, "similar to columns built by man".

What a newspaper article! And I'm sure at the time must have raised a few eyebrows, but not that many for little was said when we did land on the Moon, disappointing so many, was somebody in NASA thinking it was best to keep quiet. With the Moon landings it was always going to be secret if you found something, with so much military involvement and all the Apollo astronauts having such a background, apart from the only geologist, that being Harrison H. Schmitt, the public domain would always be the last to be considered.

The Apollo missions without a doubt were an incredible achievement. To go all the way to the Moon and back again and return alive said something about us as a race. Great courage and dedication paid off, but were we all really being given the truth. So many miles, so many new frontiers, but for whose benefit? There is one big problem with the Apollo program, apart from the cost, it was generally to show what America was capable of and if that meant keeping secrets at the same time, then why not. What was found was probably difficult to deal with as most of the world was watching, so NASA had two channels, one for the public and the other for them. In other words if they saw anything unusual that might cause a stir then communicate it on a different frequency and keep it a secret. This form of censoring out of a transmission between astronaut and M.C. is called the delayed-tape technique and allows a delay of a number of minutes before the public can hear the transmissions. The only problem NASA had was that amateur radio operators were quite capable of picking up these censored communications. One example of covering up conversations comes from the Apollo 11 mission where Neil Armstrong and Buzz Aldrin saw something very spectacular. The story is incredible but their conversation talks of them being observed from nearby hills when they were on the Moon. Otto Binder a former NASA space program member claimed that conversations had taken place on the Apollo concerning of unusual activities but outside the spacecraft, and these were of course deleted by NASA so no further questions could be asked.

Even more unusual is Dr Farouuk El Baz [NASA foremost scientist] statement back in the 70s that not every discovery had been announced and that NASA most likely used codes in some of their mission-control conversations with the astronauts. The following Apollo 16 conversation proves this point,

CAPCON: What about the albedo change in the subsurface soil? Of course you saw it first at Flagg and were probably more excited about it there. Was there any difference in it there—and Buster and Alsep and LM?

DUKE: No. Around the Alsep it was just in spots. At Plum it seemed to be everywhere. My predominant impression was that the albedo was [garble] than the fine cover on top.

CAPCOM: O.K. just a question now for you, John. When you got to halfway, or even thought it was halfway, we understand you looped around south, is that right?

YOUNG: That's affirm. We came upon-Barbera.

Im sure you would agree that this conversation was difficult to follow, but designed for certain ears only, it clearly showed the use of code when passing on delicate information. In the following conversation communications between Apollo 17 and mission control soon change to code.

LMP [lunar module pilot]: What are you learning?

CAPCOM: Hot spots on the Moon, Jack?

LMP: Where are your big anomalies? Can you summarize them quickly?

CAPCOM: Jack, we'll get that for you on the next pass.

CMP: [command module pilot]: Hey, I can see a bright spot down there on the landing site where they might have blown off some of that halo stuff.

CAPCOM: Roger. Interesting. Very- go too KILO. KILO.

CMP: Hey, its gray now and the number one extend.

CAPCOM: Roger. We got it. And we copy that it's all on the way out down there. Go to KILO on that.

CMP: Mode is going to HM. Recorder is off. Lose a little communication there, huh? Okay, there's bravo. Bravo, select OMNI. Hey, you know you'll never believe it. I'm right over the edge of Orientale. I just looked down and saw the light flash again.

CAPCOM: Roger. Understand.

CMP: Right at the end of the rille.

CAPCOM: Any chance of-?

CMP: That's on the east of Orientale.

CAPCOM: You don't suppose it could be Vostok? [this was a Russian probe]

Again you are left wondering what the heck NASA was talking about, Vostok probes had not reached the Moon yet, so to mention them at the end of the conversation was very unusual. You do feel a sense of panic as NASA suddenly realized that they might not be the only ones on the Moon or were expecting this from the start, taking it one step at a time and making sure that the public were not a part of it.

So we have looked at the use of code, which is very useful if you want to hide something. On the Apollo 16 mission, conversations and even film show us that things are not what the may seem. We'll look at that shortly. NASA has without a doubt gone to great lengths to show us all a well edited version of the Moon landings and the sad fact of this great achievement is that it seems was to benefit only a few. You do wonder why so many secrets have been made in the first place, are NASA protecting us from something? Are we not alone? Let's face it, we have all these telescopes looking for signs of life out there in the Universe and yet it could be right next to us on the Moon. That would I'm sure rattle a few nerves, never mind worry leaders, hence the keeping of it a secret.

Apollo 16 was without a doubt a mission that would reveal signs of something, dare I say life. It does seem that unusual conversations took place and even film exists, which is available to the public of a rather unusual conversation between two astronauts. It is one of those moments of the Apollo missions that really do leave you baffled. The following is that conversation.

DUKE: Look at that glass colored one right their John.

YOUNG: Pretty good size is'nt….

DUKE: Yeh.

YOUNG: look like it's about 3 days old, must be on the order of 4 billion.

DUKE: Lets lift this one up and er, ah, rat…this looks like a company on errors on a tomb.

YOUNG: Yeh.

DUKE: Hold still.

YOUNG: That is a crystal in a rock, if I've ever seen a crystal in a rock.

DUKE: First one today.

YOUNG: Yeh.

DUKE: We've got to get over this John….

YOUNG: Do you want to take off and go that way know?

DUKE: Yeh. Hey John did you make those little footprints here on this sur….?

YOUNG: Yes I guess I did yer I did.

DUKE: Ah the old footprints on the crater rim.

YOUNG: There's a good rock right there.

DUKE: Look at the size of that rock. I thought this thing was right next door to us.

YOUNG: Well there we have your half rock right there. It got black streaks coming out of it huh……

The conversation continues and we are all left wondering why footprints can be an important issue, considering we are supposed to be the only ones on the Moon. There is a big point about those words, firstly "little footprints" and secondly "Ah the old footprints on the crater rim". So excuse me when I pick out LITTLE and OLD footprints and question what must have already been on their minds to have asked the question in the first place. What that conversation does say is that something had already unsettled the astronauts so much to have asked that daft question. Lets look at another conversation that took place earlier. It really does make you wonder just what was going on at the time.

ORION: Orion has landed. I can't see how far the [garble]….. this is a blocked field we're in from the south ray-tremendous difference in the albedo. I just get the feeling that these rocks may have come from somewhere else. Everywhere we saw the ground, which is about the whole sunlit side, you had the same delineation the Apollo 15 photography showed on Hadley, Delta and Radley Mountains....

CAPCON: O.K. Go ahead.

ORION: I'm looking out here at Stone Mountain and it's got- it looks like somebody has been out there plowing across the side of it. The beaches – the benches – look like one sort of terrace after another, right up the side. They sort of follow the contour of it right around.

CAPCON: Any difference in the terraces?

ORION: No, Tony. Not that I could tell from here. These terraces could be raised but of [garble] or something like that.....

CASPER: [**lunar orbiter overhead**]**:** Another strange sight over here. It looks – a flashing light- I think it's Annbell. Another crater here looks as though it's flooded except that this same material seems to run up on the outside. You can see a definite patch of this stuff that's run down inside. And that material lays or has been structured on top of it, but it lays on top of things that are outside and higher. It's a very strange operation.

NASA of course says that all words used in this conversation are just metaphoric terms. The problem we have is that those words can and do tell another story.

Quite frankly to go all the way to the Moon and then use known words so clearly and say they are metaphoric terms does not seem believable. You have clear descriptions, "it looks like somebody has been out there ploughing", and of course, "Hey John did you make those little footprints here on this sur...". What ever you want to make of these sentences they speak for themselves and tell us something. We seem to have gone all the way to the Moon and abused the English language.

As we have discussed earlier secret code, different channels and now clear language but NASA version. So does it mean exactly what they are saying? Going to the Moon was without any doubt a great achievement and those men were very brave, but they were well educated and trained, so all their sentences were spoken with intelligence and truth. They saw things which were possibly known about already, that were not natural and were not that old and belonged to some one else.

You really do have to give credit to those men who went to the Moon, not only was the job dangerous but they had to lie as well. Somebody out there wanted secrets to stay that way, and with the military involved it was not worth antagonising them.

So Apollo and all its missions found something, whether they expected this or it was just a big surprise does not really matter, because they did not tell you, the people. The Moon had already been visited, and there were clear indications of that and I'm almost certain there is still somebody there. The following conversation from the Apollo 15 mission, talks about tracks.

SCOTT: Arrowhead really runs east to west.

MISSION CONTROL: Roger, we copy.

IRWIN: Tracks here as we go down slope.

MC: Just follow the tracks, huh?

IRWIN: Right we'er [garble]. We know that's a fairly good run. We'er bearing 320, hitting range for 413….. I can't get over those lineations, that layering on Mount Hadley.

SCOTT: I can't either. That's really spectacular.

IRWIN: They sure look beautiful.

SCOTT: Talk about organization.

IRWIN: That's the most organized structure I've ever seen!

SCOTT: It's [garble] so uniform in width.

IRWIN: Nothings we've seen before this has shown such thickness from the top of the tracks to the bottom.

An amazing conversation. Clearly both astronauts had found something and that was that somebody else had been there before them! This whole business of tracks on the Moon goes on and on, with photographs of certain objects actually moving across the Moon's surface. Ref [PLATE 23 NASA PHOTO LO V, NO 67-H-1135] and [PLATE 24 APOLLO 16 PHOTO, NO 16-19067]. It must be made clear that both these pictures on an official level show a meteorite skimming along the surface of the Moon, as far as most astronomers are concerned. Although I can see their point it does look more like a tracked vehicle than a meteorite. There are too many descriptions that sound more like an intelligence being involved, than all of this just being natural, even the astronauts point out the bloody obvious!

What should also be brought back to our attention is what Major Patrick Powers said [Head of the United States Army Space Development programme] back in 1968, " the first man to reach the Moon must be prepared to fight for the privilege of landing". A statement like that in itself tells us that someone else is on the Moon, and if that was not enough Doctor Carl Sagan was already convinced that the Moon had had, or still had occupants with plenty of signs they were there. How much evidence has been covered up? It seems the astronauts were clearly looking at evidence of life, as so many conversations were either being blocked or were obviously about life! Lets not forget that NASA is a civilian agency but part of the money for projects comes from the Department of Defense. Most astronauts are military officers subject to military regulations. This in turn connects us to the NSA [National Security Agency], which monitors all the transmissions and looks over all the footage from missions. Information is almost zero to the public and if you want to keep a secret then the NSA knows how.

So back to the Apollo missions and even more unusual conversations as the following reveals on the Apollo 17 mission,

MS [Mission Control]: Go ahead, Ron.

EVANS: OK, Roberts. I guess the big thing to want to report from the back side on this one is that I took another look at –the-cloverleaf in Aitken with the binocs. And that southern dome [garble] to the east.

MS: We copy that, Ron. Is there any difference in the color of the dome and the Mare Aitken there?

EVANS: Yes, there is…That Condor, Condorsey, or Condorecet or whatever you want to call it there. Condorecet Hotel is the one that has got the diamond shaped fill down in the uh-floor.

MS: Robert. Understand. Condorcet Hotel.

EVANS: Condor. Condorcet Alpha. They've either caught a landslide on it or its got a- and it doesn't look like [garble] in the other side of the wall in the northwest side.

MS: OK., we copy that Northwest wall of Condorcet A.

EVANS: The area is oval or elliptical in shape. Of course, the ellipse is toward the top.

That conversation was about Domes, which back in the 1950s seemed to be popping up and disappearing in all sorts of places on the Moon. I would rather leave the subject of Domes alone, as that subject on its own could fill a book and create many more arguments. Domes it seems in most astronomers eyes are a mystery but one I find very much one sided when conclusions are made.

The Moon has some very big secrets that are being shared with only a few, but what does interest me is that who ever is up there, does not seem to be a threat. The Moon has had probes from both the USA and Russia. There are reports documented and filed out of sight that indicate that we were well monitored in space, but never threatened. The Moon is strategically positioned, ideal as an observation post for looking at us. The other side of the Moon of course could hide many secrets and that is the most likely place to find intelligent life as quoted by Carl Sagan in 1962 and Dr. Walter Riedel the late director of the Peenemunde Base in Germany.

With the abundance of materials on the Moon what's to say it's not already being mined? There's more than enough evidence to back that up. Back in the 90s I came across a picture taken on two separate occasions from different orbitors in different years, of one crater. What surprised me was that on the second picture there was something that was not there when it was originally photographed on the first and it looked like a crane involved in some form of mining work. Continuing on this subject west of the crater Aristarchus, [LUNAR ORBITOR IV Photo no. HR 157], the picture shows a circular crater next to an almost triangular entrance. Do you not get the feeling that NASA only gives us half the story? Whenever somebody actually finds something in space, on the Moon, Mars etc. you feel they have been briefed to tell you bits of information and the rest of that information stays secret. One of the major changes in the space race since the Soviet Union disbanded, is that the new Russia has been very open in areas of space exploration and has released documents and pictures of very unusual encounters in space. I'm not saying everybody in NASA is part of a huge conspiracy to fool us all, but that something is wrong, when you look at the evidence I have shown, is not really explained. The whole set up is wrong, it seems everything is said to be natural if the final conclusion points to life and clearly contradicts the accepted protocols.

The fact is that NASA has official photographs of a Lunar base which was photographed by the Lunar Orbiter and filmed by Apollo astronauts and to add to that photographs of structures, tracked vehicles and so on. On this occasion use your imagination and you will be correct. We have already had an insight into how good the covering up of information goes, but at the same time I found holes. These holes become a jigsaw and when we join them together it seems to show the whole space program is an expensive cover up.

It's not fair of course on those astronomers who have tried to make a point but have been shunned and put into the corner by other so called astronomers. Astronomer Gruithuisen's puzzle of a Lunar fortress which created a huge debate at the time and Professor William H. Pickering's observation of a snow storm on Mount Pico [PLATE xxxiib], which would point to some kind of atmosphere. Pickering's observations on their own were very valuable, one of those observations being south of the Mare Imbrium where you can find the crater Eratosthenes. Pickering repeatedly observed grayish spots moving around inside the crater. The many theories put forward were not that daft either, ranging from clouds and even vegetation.

It would sound daft to most people to say there was vegetation on the Moon, but it does seem that something very similar had been found. I must remind everybody reading this book that you only see one side of the Moon. It is suggested and without revealing names that the other side of the Moon has more activity on it than you could ever imagine. There are again some very unusual photographs out there which do seem to show vegetation.

NASA and the Russians have these pictures but are not willing to share them because they know exactly what the implications are.

In the next chapter colonizing the Moon, are we there already? Is someone else there? Or are we already in contact?

Chapter 3

Life On The Moon

THE END OF 2006 LEFT US with news of the return to the Moon. It seemed after all those years we finally were going back and in years to come live there and use it as a base to travel to Mars. I must say how that would be possible puzzles me, when we are supposed to be there [already or somebody else]. The most difficult thing for the NSA [National Security Agency] will be other countries, intentions to visit the Moon. You have India, China, Japan and the Europeans planning all sorts of pioneering trips to the Moon and they will include human beings. Whatever the agreement made, if anything was agreed, it will be interesting to see how they pull it off. To be fair the Moon is a big place and we are only shown small areas, so if you want to hide something it should not be a problem. Perhaps everybody is after a part of the Moon, big money to be made if you can use its endless resources.

Back in the 80s it seems NASA had big plans for colonizing the Moon. In 1986, Alamos National Laboratories unveiled the SUBSELENE a device for tunneling under the Moon's surface. The machine would be in the shape of a rocket, using a nuclear reactor. The idea was that heat produced from the reactor would melt rock into glass-walled tunnels, which would be used for high-speed transportation. Because the SUBSELENE could produce glass-like material, this could also be used for making bricks for construction on the Moon's surface. If you don't mind the technical jargon then this is how this device would work.

The nuclear power source would come from a very small SP-100 fission reactor. This would deliver heat to the tunneler head from liquid metal pipes, using lithium. Heat produced would not only provide a way of melting the soil and rock, but also power the mechanics and electrics. With cutting diameters of 3m and 5m, 260ft a day would be possible. Two devices were planned using the SP-100 fission reactor. After arriving and landing on the Moon, the device would bore to a depth of about 65ft. From this point it would level out and create a tunnel. When all tunneling was complete, manned crews could take over and anything could be created in the tunnels, from mining to scientific study to even agriculture. Of course it would be difficult to say if this device ever left Earth. It was patented in 1975 and possibly stayed on the drawing board. But there is something going on and as we have discussed in other chapters anything is possible, and that brings us to the big question. Did we still keep going to the Moon? I rather feel tempted by that phrase, "The show must go on". That show of course could have been the whole Apollo program in its well edited version screened to us.

Now time for a change, what I have found really fascinating over the years, are the number of books published that mention vegetation on the Moon. Most people seem to laugh at any suggestion that the Moon is more than just a silent body and that anything is growing on it. Going back in time to the 1900s a well respected astronomer W.H. Pickering had observed very strange dark areas that moved every monthly Moon cycle. The observations were seen especially in the crater Eratosthenes. Now what Pickering did next was quite extraordinary, he put his reputation on the line,

"It is perfectly obvious that terrestrial vegetation could not exist on the Moon and probably not on Mars. But something does nevertheless exist on both of these bodies, which, while differing more or less from anything with which we are familiar, can be better described by the word vegetation than any other in the English language."

Pickering had without a doubt taken a risk, but was certain that what he saw was what he had concluded.

This reminds me of the Apollo 8 photographs which were taken on the Lunar backside [REF PLATE44,45]. What they seem to show is the Moon in Autumn colors, sounds mad, but what are we looking at? Vegetation. It's even more interesting to read William Coopers [Ex US Navy] comments in his book- Behold A Pale Horse. He talks about areas on the Moon that change in color and vegetation growth. He goes further and refers to man made lakes and even clouds, something we have already discussed. Cooper even says he has the original photographs taken by NASA. It is very difficult to imagine a Moon that for so long has been portrayed as dead having some earth-like qualities.

We seem to just take for granted what we are told, and if it does not directly affect us, then never question it. I find it very difficult when astronomers and scientists say "Not possible" and the next moment they are wrong. It seems there is a lot of evidence to say the Moon is not as dead as we are made to believe. At one time and in many astronomers books it was said water did not exist on the Moon and of course they were wrong. It does seem that information is released a bit at a time, and by the time the public have got used to new information about the solar system, the silent mode comes in for a long time. What I am trying to say is we get very small pieces of information that do not really create much interest, the big stuff is top secret. Let's put a few opening questions. What if we have a base already? - Have we already found intelligent life? - Are we jointly sharing the Moon? -Who is in charge? -Why all the secrets? - How far do you go with lies? - Where are all the benefits to mankind?- Is the technology from somewhere else?

In general all the above questions should be easily answered and apply. It's rather interesting when you look back at NASA and its general attitude towards the Apollo program. It was criticized for failing to provide scientists on early Apollo missions. It was not till after Apollo 14 arrived in Lunar orbit that comments were made by Eugene Shoemaker [a leading figure in lunar science], who at a later news conference told NASA what a miserable job they were doing and that it was a waste of 24 billion dollars. NASA knew of Shoemaker and his general criticisms and of others in the scientific community and had a plan to focus more on the Moon's geology in the next three missions. As we know on the Apollo 17 mission the first geologist to walk on the Moon, was Harrison Hagan Schmitt. He may have been on the last mission, but his geology skills paved the way for many interesting finds.

What we must not forget is that, there was going to be an Apollo 18 and 19, and these missions would have explored areas of the Moon which would have been of great scientific interest. Due to lack of funding and public boredom the Apollo program was cancelled. You do feel though that perhaps NASA was leaned on by powers unknown because they were getting a bit too keen to explore further on the Moon.

It is interesting to read in a number of books that the Americans have a Moon base shared with Russia. There is no real evidence to confirm this, but when certain sources were questioned over this matter, the probability of this being true was 85%. What was even more of a surprise was that we we were not alone, but sharing with the Moon non-humans. I must say that it does not surprise me to hear this. When you look at all the unusual conversations between astronauts and pictures, which show strange objects you can easily come to this conclusion.

This brings me to more of the unusual photographs which are available to the public, and which are odd. We enter the business of mining again and bridges and weird symbols and even letters.

One of the most interesting pictures I have come across is a photograph of a crater [not named] that seems to show a white cross within it,[PLATE 88, APOLLO 8]. What is even more interesting are the unusual terraces, which look as though they have been cut with some sort of machine. To go further there seems to be a very strange black mark running central to the crater, just like a bridge. When you think back to Dr H.P. Wilkins at the Mount Wilson Observatory, USA, and his observation of a 12 mile long bridge elsewhere, it does not seem that crazy to suggest this particular crater could have one as well. This picture could easily be taken for some mining operation on Earth, look at the similar markings and almost road like tracks and with the reports from astronauts about tracks being seen on the surface, as well as moving vehicles being photographed, you do have a lot of evidence to back up the possibilities.

I must point out that I am only talking about a small number of pictures here and these are the hardest to explain, as NASA and the NSA has gone to great lengths to make sure that what is released will not give anything away. Remember what they actually have, would change everything about us as a race and I'm sure, shock a lot of people.

It's always fascinated me that a number of pictures you come across of the Moon's surface show letters on them. It is one of those moments where I feel like laughing, we go all the way to the Moon and find not dust but plenty of letters. What does fascinate me is the fact that most of these letters are from our own languages. You wonder if the areas are marked so from an aerial view that way you can locate positions easily. Here on planet Earth it has always been a good way of locating things, people etc, by leaving marks on the ground, that can be seen from the air. Of course it all could be natural, but I would not be writing this book if I thought it was that simple. It is very easy to say," you are wrong" and that everything I have talked about is nothing but complete fantasy, but that's the easy option! Those in charge will continue to pull the wool over your eyes if they can get away with keeping secrets. The fact that I am now writing this book, shows that leaks are becoming common and getting more detailed.

I remember years ago somebody saying to me that the Apollo Moon landings were all filmed on Earth and that it could be proved through the position of light on the astronauts and dodgy photographs. I do not really want to get involved in that debate, everybody is entitled to their own conclusions. What I would like to point out to those doubters, is that perhaps it was a cover story put out to take you away from what was really going on. Please remember that this is my own theory, although it is worth thinking about.

If NASA spoke in code and on different channels and kept photographs from us they had a massive reason. Colonizing the Moon is it seems the intention of the Americans anyway, so whatever the outcome some secrets will have to become public. It was not that long ago when people like Richard Hoagland were arguing about glass structures on the Moon [and I must say with a lot of evidence]. He showed photographs and pointed out the usual problem of missing pictures. NASA it seemed was very good at misplacing important photographs when you needed them. When you consider how many were actually taken and how many we actually have been shown, you do wonder what are on the others, which are either locked away or have been destroyed.

I personally think that Richard Hoagland brought an amazing insight of the possibility of glass structures on the Moon, because glass would be far stronger than certain metals in space. On Earth people experiment with the glass structure theme, living for months within them to find out if it is possible to create a habitable environment. You know it is so possible that this is all happening at this moment on the Moon. It does seem that 1972 was not the last time we walked on the Moon, nor 1969 the first time. Everything I've talked about does not point to a silent Moon and points more to a very busy, but secret place. We are only showed what is acceptable, as the unacceptable is the truth.

Chapter 4
The Conclusion and Weird Words

WELL I HOPE I HAVE OPENED your minds and given you a new outlook on our very special neighbor. I completely believe that we are not alone out there and we would be daft to presume we were. The Moon is so close and yet, beyond reach. It has been described as a spaceship or an object created from our own Earth in the beginning. Whatever you would like to believe it is special and always with us. I look up into the sky sometimes and find it just sends you into a trance, as you wonder at its beauty and mystery.

It is hard to believe that it is alive, for so many out there continue to disbelieve and ridicule those who dare to challenge the old and established system. How many people does it have to take before the sceptics listen? Look at the many people in our past who have challenged our ways of thinking, and who were either were mocked or even executed for their beliefs. Like I have said from the start, all options should be left open and discussed. The human race is very good at concealing the truth, and for some reason will go to great lengths to cover up facts, and that, I'm afraid, is exactly what has been done in the case of our neighbor the Moon.

For the first time I will mention UFO [Unidentified Flying Object]. Back in the 90s it was one of my favorite subjects and one which was to lead me down all sorts of unusual avenues. I researched all kinds of sightings and was friends with many others following the same trails and reaching the same conclusions. Those conclusions without a doubt told us we were not alone and that we were well monitored.

I remember a story about a leading figure in government in the UK, who was having a difficult time with some sightings that had occurred. He was so upset about what he knew that on one particular occasion he broke down in front of his friends. What he said was very shocking, that we were not alone! What he actually said was that, our air space was invaded all the time by unknowns and there was nothing we could do about it. It is worth remembering that if you do report anything to the MOD they usually reply by stating that what you saw, "is of no defense significance" and with a statement like that you are left speechless. This could easily lead you down the road to wonder about Military Projects, but alas the technology is far superior and Military Projects are not usually carried out over housing estates.

So why am I telling you all of this? The answer is very simple when we went to the Moon we were monitored on every occasion and some astronauts have admitted this. Photographs were taken and even film exists with encounters of UFOs. There are many television programs that cover this subject now and hopefully they will go further and investigate the real facts about the Moon. We need to know the true facts as secrets only become bigger in the long run while we try to uncover the truth.

Years ago Neil Armstrong narrated a series, which was about the history of aircraft. It was very well presented and gave you a great insight into our achievements in flight. What I found really interesting was his description of the Lockheed F117 stealth bomber, which he referred to as alien-looking. I know it could easily be an innocent remark, but you almost feel he had seen something that was alien to be able to give this comparison. Neil Armstrong is one of those astronauts described as seeing things on the Moon that were of unknown origin.

Once again I would like to point out that nearly all astronauts are connected with the military, so are under strict rules that do take away many of their rights. So it is very difficult for them to reveal the truth under the watchful eye of their silencers.

It all sounds rather sinister, but these areas of secrecy are very murky. If we have already made contact then we have already gone to the stars. Technology is accelerating so fast, you wonder if some of it has been obtained through cooperation with others. Another situation I heard about was that of an astronaut walking past a hanger and being quite shocked to see some form of spaceship that did not need to sit on the end of a rocket. He was later to complain about why they were not using the technology he had seen. Of course whatever he saw was most likely not for the public domain.

That brings us back to the Moon, which is so close and so Two Faced that you can get away with just about anything on one side of it. All the facts connect together somewhere along the line and the closer to the source you get the more you feel the pressure. It is ridiculous to think that when the Apollo astronauts were going to the Moon, there was a technology out there far more efficient and safe and we were using it jointly with alien beings. Most people will probably laugh at this [and I don't blame you] for doing so, but if it is true, then we have some serious problems.

I would once again like to remind you of what prominent people said in the past about the Moon,

1. Major Patrick Powers, "the first man to reach the Moon must be prepared to fight for the privilege of landing".
2. Dr Carl Sagan, " mankind must face the probability that beings from elsewhere in the Universe have, or have had, bases on the averted

side of the Moon".
3. Dr Farouuk El Baz, "not every discovery had been announced".
4. Dr Walter Riedel, who believed that intelligent life was already on the Moon.
5. Otto Binder, claimed unusual conversations had taken place between astronauts while on the Moon.

To follow on from Otto Binder and the conversations, let's remind ourselves of some very unusual ones.
1. Apollo 16 and astronaut Duke says the following, "Yeh. Hey John did you make those little footprints on this sur...?"
2. Apollo 15 and again strange tracks this time. Irwin, " Tracks here as we go down slope".

It is difficult to believe that all these people were just saying things just for fun. All concerned were in prominent positions at the time and no doubts knew exactly what they were talking about. It is therefore a shame that in later years some of the above joined the club of keepers-of-the-secret.

1969 and Neil Armstrong says those famous words, "That's one small step for man, one giant leap for mankind". If only those words were true. You could say it was one giant leap backwards especially in keeping secrets and if it meant silencing astronauts for life then so be it.

It is interesting to read how going to the Moon changed their own lives. They all were never to forget how another world could cause so many different feelings to be brought back with you. I will conclude with what I believe is going on, with regards to the Moon. The evidence clearly points to a Moon already explored and of both lost and current civilisations, including us. This means we have already made an agreement with somebody not of our species. They probably were not willing to join us, but had to due to mishaps, which brought us together. That's where technologies became available to us to expand our knowledge. If you think this sounds crazy then look carefully at how we have evolved in such a short period of time in the area of technology. And I can tell you that you don't even get to see most of it, as secrecy rules, especially within governments.

We are already on the Moon and have been there for a long time, trying to get on with others and joining in on the mining operations that have been going on since we landed. The moon is a world of valuable materials and a fantastic watch post on keeping an eye for us. There will come a day when as a world we will be united, and it may take somebody from another world to achieve it. We may even grow up, but it will be along time yet before "THE TWO FACED MOON" gives all its secrets away.

Points to Remember

I would like to point out that on many occasions I have requested certain pictures taken either on or in orbit around the Moon from NASA and either get a childs starter pack on the Moon or no reply at all. Some departments were helpful, but did not carry the information I required. Many pictures that I have studied have been sent to prominent people for their opinion and all replied. What was interesting was how they seemed more baffled than I was at what I had sent them and would follow with, "keep up the good work" There are many people out there following the road I decided to take and I wish them all good luck, and hope we can crack all the secrets in the end.

The Apollo Astronauts

Apollo 7-1968 Walter Shirra, Donn F. Eisele, R. Walter Cunningham.[first Earth orbit]

Apollo 8-1968 Frank Boreman, James A. Lovell, Jr., William A. Anders. [first Moon orbit]

Apollo 9-1969 James A. McDivitt, David R. Scott, Russell L. Schweickart. [Earth orbit test of all modules]

Apollo 10-1969 Thomas P. Stafford, John W. Young, Eugene A. Cernan. [practice landind for lunar landing]

Apollo 11-1969 Neil A. Armstrong, Micheal Collins, Edwin Aldrin, Jr. [first Moon landind]

Apollo 12-1969 Charles Conrad, Jr., Richard F. Gordon, Jr., Alan L. Bean [second Moon landing]

Apollo13-1970 James A. Lovell, Jr., John L. Swigert, Jr., Fred W. Haise, Jr. [failed Moon mission]

Apollo 14-1971 Alan B. Shepherd, Jr., Stuart A. Roosa, Edgar D. Mitchell. [first scientific visit to the Moon]

Apollo 15-1971 David R. Scott, Alfred M. Worden, James B. Irwin. [first extended mission]

Apollo 16-1972 John W. Young, T. Kenneth Mattingly II, Charles m. Duke. [first visit to Moon's central highlands]

Apollo 17-1972 Eugene A. Cernan, Ronald E. Evans, Harrison H. Schmitt. [final visit to the Moon and the first professional scientist]

Apollo 18- CANCELLED

Apollo 19-CANCELLED

Thank You's

All though I carried out my own study on the Moon the following books were used as ref,

COSMOS—Carl Sagon

BEHOLD A PALE HORSE—William Cooper

GUIDE TO THE MOON—Patrick Moore

A MAN ON THE MOON—Andrew Chalkin

THE WORLD OF THE MOON—Henry King

OUR MYSTERIOUS SPACESHIP MOON—Don Wilson

MYSTERIOUS VISITORS—Brinsley Le Poer Trench

THE CONQUEST OF SPACE—Bonestell—Ley

THE MOON BOOK—Bevan M. French

A FIRE ON THE MOON—Mailer

THE NATURE OF THE UNIVERSE—Fred Hoyle

THE INVASION OF THE MOON 1969—Peter Ryan

FIRST ON THE MOON—Micheal Joseph

SOMEBODY ELSE IS ON THE MOON—George H. Leonard

All pictures used are from the archives of NASA. Pictures taken from film were carried out myself. All reports were purchased and original newspaper cuttings used as necessary.

Lockheed F117 Stealth

The eye in the sky

Moon secrets soon to be explained
Human landings may be challenged

American and Russian moon probes are preparing for the first extraterrestrial landing by humans. Will man be allowed to land? Readers who have seen the film "2001: A Space Odyssey" will be familiar with the speculations put forward in this article. They may be surprised to learn, however, that the film story derives much of its inspiration from observations recorded over the past 120 years or so.

The hidden side of the moon taken by Lunar Orbiter III at an altitude of about 890 miles.

By Paul Nicholson

SOME time next year the first lunar landing will be attempted and sooner should there be forthcoming to the photographs of the Moon's surface that have puzzled astronomers for years, strange things that have even led some eminent researchers to speculate that our natural satellite may already be occupied by an alien intelligence who are using it as a base.

Major Patrick Powers, head of the United States Army Space Development Programme, has said that "the first man to walk the Moon must be prepared to fight for the privilege of landing". Unfortunately Major Powers, writing in a magazine, failed to amplify this theme, thus leaving us with an interesting but enigmatic statement.

In December 1962, at the convention of the American Rocket Society in Los Angeles, Dr Carl Sagan, an adviser on extraterrestrial life to the US armed forces, said mankind must face the probability that beings from elsewhere in the universe have, or have had, bases on the reverse side of the Moon.

This possibility demands serious consideration on the basis of negative evidence gathered by astronomers over the past 150 years.

Astronomers began reporting strange lunar activities in the early 19th century. On October 20, 1824, lights, flashing intermittently for half an hour, were seen on the darkened part of the Moon. Four years later several astronomers witnessed many bits independently, not a display of flashing dots and dashes that moved about in the Mare Crisium.

In February 1836, two straight lines of light, with luminous dots arranged symmetrically between them, took over the following year the Lunian astronographer Maller surprised astronomers by producing a Moon map which included an oblong on the edge of the Mare Frigoris. Sixty-five miles long, the oblong had a perfect white cross in its centre. The tracing was verified by another astronographer, Nelson, who described the thing as between 276 and 170 feet high.

Optical illusion?

Many lunar oddities were reported by a Copenhagen observatory in 1842. Two luminous triangles on the Moon's upper limb were sighted. These features were not only strange but appeared on the lower limb. They rapidly approached each other, met, and then disappeared.

Seven years later, a huge equilateral triangle showed up on the Crater Plato and tiny points of light were seen all over the Moon, staggering in different orders and converging on Plato.

Plato, in fact, seems to be a spot of great activity. In 1871 an astronomer called Birt gave the Royal Astronomical Society a list of some 1,600 observations he had made of light changes, moving objects, flashing lights, and globe-like patterns inside the lunar crater.

Lunar anomalies, however, are in no way confined to the 19th century. In the early 1950s astronomers discovered a new brand of minor puzzles, in the forms of rounded white domes.

These may, we know as Moon's interior to be active. Many of the sightings of strange lights may no doubt be rationally explained as outgassings.

On the night of November 26, 1956, Robert Curtis, a New Mexico astronomer, who went by pictures of the Moon with a Fresca-type camera fixed to his six-inch reflector telescope. When the film was processed Mr Curtis had a shock. To the left of the terminator (the sunlight-shadow dividing line) was a white Maltese cross. It appeared to frame after frame and was obviously on or just above the Moon's surface. Due as far as is known, has not been satisfactorily explained.

It was the late John O'Neill, former science editor of the New York Herald Tribune, who really caused the biggest stir in Lunar observations when, in July 1953, he reported what he called a "gigantic artificial" with 12 miles long—the Lunar Bridge.

Controversy

The finding caused immediate controversy, fully ended by Dr H. P. Wilkins who, using a 54-inch reflector at Mount Wilson Observatory, USA, confirmed the Bridge. He, however, believed it to be much less than 12 miles long and a natural phenomenon. Strange, then, that a month later, when viewing conditions were again ideal, the Bridge had disappeared.

A year later Frank Halstead, an eminent astronomer, his assistant, and 16 visitors, observed a straight black line on the base of the crater Piccolomini.

When all this was going on, most people had forgotten about Moon quests. But they were to make a reappearance in 1966, after many failures, the American Ranger Seven spacecraft succeeded in with the first close-up pictures of the Lunar surface. Some 4,200 were taken. Two showed small white domes, 7½ x ten diameter at eleven feet directly outside.

On November 18, 1966, a lunar Orbiter probe left Cape Kennedy. Three days later, from a height of 30 miles, the spacecraft's telephoto lens was focused on the Moon's surface and the camera activated. On a small section of the Sea of Tranquillity, a plain just off the Moon's centre, the cameras picked up six spike-like projections, casting shadows across the dusty surface.

They were called some of the most unusual features of the Moon ever photographed by the scientists in charge of the mission. However, they left like spikes to be natural. Mr William Blair, a lunar anthropologist and a member of the Boeing Company's biotechnology unit, thought otherwise. They were, he maintained, in a geometric pattern, "similar to columns built by man".

The scientists are still wrangling about such mysterious features. Meanwhile, the Moon remains a big puzzle. Our small satellite, only 2,160 miles in diameter holds its secrets well, or Lord them. Those perhaps we will learn the answers to the strange goings on that have terrified us for a century and shall have simply before we are in for a big shock.

SCIENTISTS ARE PLANNING TO MINE

Moondust as the new

By Roger Todd

ONE small step on to the Moon's surface by Neil Armstrong could yet bring the Earth untold riches. While poets, songwriters and astronauts wax lyrical about moonlight, moonbeams and moon travel, it has been feet-on-the-ground scientists who have heard the penny drop.

With man digging and drilling deeper and deeper beneath the Earth's surface for diminishing supplies of oil and coal, physicists have discovered a fuel of the future: "Moonglow."

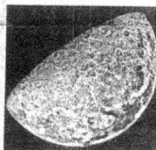

RICHES: Bright side of the Moon

There is enough fuel lying in the first 10ft of the Moon's surface to power the Earth for the next 700 years.

Now the scientists are desperate to send miners into space to dig it up and transport it back by rocket.

Operation Moonglow is a venture exciting enough to make Jules Verne eat his heart out with envy.

The scientists have already mapped out how it can be done with international co-operation.

Their efforts are all the more pertinent as Britain last week announced the scrapping of two more planned nuclear power stations — and as more of our coal mines are shut down.

The wonder fuel lying buried in the Moon's soil is Helium-3.

The U.S. currently produces 20 kilogrammes of the substance each year as a by-product of its nuclear programme.

It costs about £300 million a ton — 35 times more than the price of gold.

One of the senior physicists working on the project is Dr John Santarius, of Wisconsin University.

He said: "We found out about this amazing fuel from samples brought back by the Apollo programme sending men to the Moon.

"Scientists have long been trying to find a clean alternative to radioactive nuclear power and this is it.

"You don't need a lot of it to produce a great deal of electricity. Just 25 tons would power the United States for a year — and that could fit into a single space shuttle bay."

IN fact, the Moon has about a million tons of Helium-3 in deposits of Titanium, a metal which has soaked it up from the solar wind. Dr Santarius said: "The problem is getting it back from the Moon. First we need to get astronauts, robots and digging equipment up there.

"The 'miners' would be 10-metre wide bucket wheel excavators, with big shovels like on a Ferris wheel, which would dump the lunar soil on to a conveyor belt.

"From there it goes to an enclosed shelter where it is heated up by the sun's rays focused by an orbiting mirror. The lunar day lasts for 14 of ours and the daily temperature is 127 degrees Centigrade.

"The mirror would heat it up to 700 degrees and more or less boil off the Helium-3.

"Then, during the 14 days of darkness, when the temperature plummets to minus 173 degrees, you condense the gas and put it into oxygen-style cylinders for shipping back to Earth."

"What is required to make the plan take off is an oil crisis, when everyone will be searching for alternatives.

"We believe that we can get the cost of power down to the current level of electricity production.

"At present it would cost 10 per cent more. But we could end up at the same price as now."

THE effects of "opencast" mining on the Moon's surface have been carefully considered. The face of the "Man in the Moon" would not be affected.

Scientists have even worked out how to avoid changing the Moon's reflectivity — because they don't want to put a stop to romantic moonlit nights.

Dr Santarius says it is simply a matter of not digging deep enough to create new craters in the lunar seas where the Helium-3 is concentrated.

The Japanese have advanced plans to mine the Moon by the year 2015. Europe aims to get there by 2020 and robots are to be sent before the end of the century.

Prof Hiromu Momota, at Japan's National Institute of Fusion Science, believes that using the Moon's vast store of Helium-3 will also make it possible to build machines to power spacecraft and even generate high-energy beams to treat cancer.

The Moon, he said, has enough Helium-3 to power the world for centuries. And after that? "Jupiter and Saturn have almost infinite reserves," he added.

International co-operation to find new fuel sources is well advanced. Britain, the U.S., Europe, Russia and Japan have already linked up to produce electric power through fusion, rather than radioactive nuclear fission.

At the Joint European Torus laboratory, in Culham, Oxfordshire, plans to use fusion are well advanced.

A spokesman said: "Fusion is about 10,000 times safer than nuclear power."

He added: "Helium-3 is horrendously rare. But if we can get it from the Moon it will be the preferred fuel."

'Just 25 tons of Helium-3 could power all of the U.S. for a year'

When someone anonymously released to Richard Hoagland, head of the Enterprise Mission (Face on Mars research) photographs of the lunar surface claimed to have been taken by the Clementine space probe less than a couple of years ago, a storm of controversy broke loose. The Clementine images appeared to show anomalous artificial features on the Moon which defied belief, and for some, Richard Hoagland's credibility gap had passed the point of no return. The Clementine probe was itself perplexing to some, given that the US Department of Defense had unrestricted and exclusive access to data for six months after NASA scientists had conducted their experiments.

Had Richard Hoagland announced that the Clementine probe's lasers discovered ice on the Moon, the man would have been dressed 'certifiable' in some quarters, but on Tuesday, 3 December 1996, that's exactly what the Pentagon announced had been discovered in a Washington DC press briefing. The ice was found in a huge crater deep in the south pole of the Moon, according to Rick Lehner, spokesman for the Pentagon's Ballistic Missile Defence Organisation. He said that a panel of scientists had concluded that the ice is frozen water rather than methane or ammonia.

"It's an extremely significant discovery," said Lehner, adding that water would make exploration easier. With water there you could have enclosed areas to grow plants, grow your own food, make your own fuel, make your own air," he said. "You don't have to launch all that stuff from big rockets on Earth."

Anthony Cook, astronomical observer at the Griffith Observatory in Los Angeles said, " If you could wish for any one thing there to make it easier to explore with, it would be water."

Lehner said the crater is twice the size of Puerto Rico and 13 kilometres deep, or as high as Mount Everest, the tallest peak on Earth. He said the ice formation is the size of a small lake and between 10 and 100 feet deep.

"People have theorized that there may be water on the Moon, but the (six) Apollo missions didn't find any evidence," said Lehner. Scientists believe that about 3.6 billion years ago, a comet crashed into the Moon, and water droplets on its tail were left in the bottom of the crater, the deepest hole in the Solar System. Because the dark side of the Moon is always dark, the temperatures in this crater are near

ICE ON THE MOON?
ANNOUNCED 3rd DECEMBER 1996

NASA TECHNICAL REPORT

NASA TR R-277

CHRONOLOGICAL CATALOG OF REPORTED LUNAR EVENTS

by

Barbara M. Middlehurst
University of Arizona

Jaylee M. Burley
Goddard Space Flight Center

Patrick Moore
Armagh Planetarium

and

Barbara L. Welther
Smithsonian Astrophysical Observatory

NATIONAL AERONAUTICS AND SPACE ADMINISTRATION • WASHINGTON, D. C. • JULY 1968

REPRODUCED BY
NATIONAL TECHNICAL
INFORMATION SERVICE
U.S. DEPARTMENT OF COMMERCE
SPRINGFIELD, VA. 22161

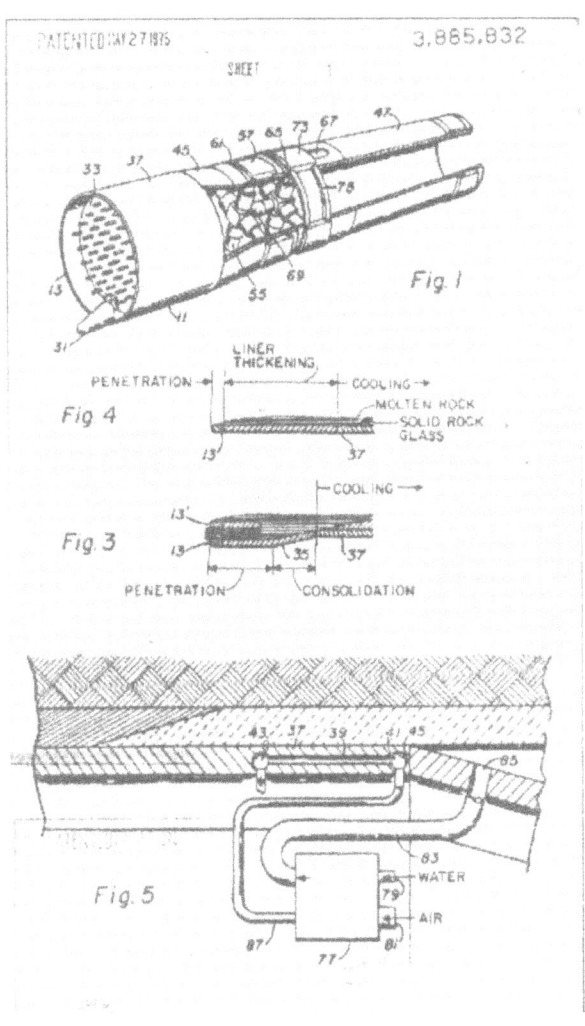

Patented May 27 1975 SUBSELENE SP-100
(Designed for landing on the moon).

*Lunar Orbiter IV photo NO HR 157 (West of Aristarchus)
(A very unusual picture, moon base perhaps?).*

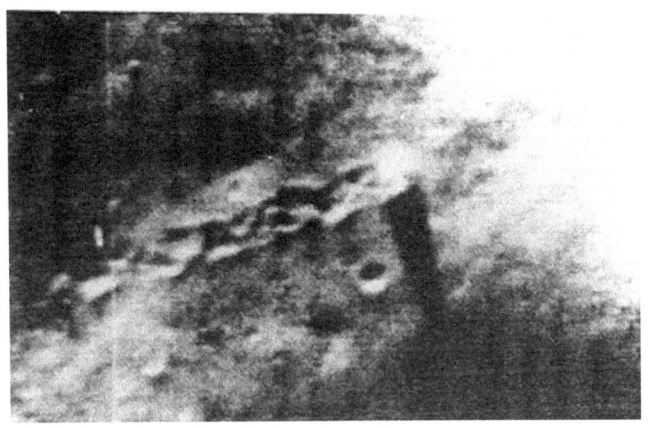

Plate 23 (NASA photo LO V)
Tracked vehicle moving along Moon's surface.

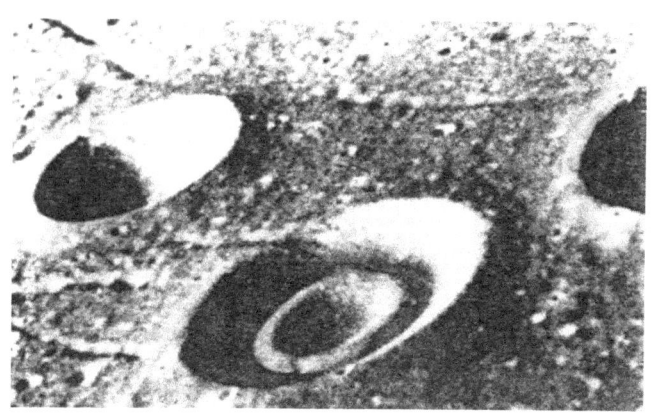

APOLLO 15 NO 15-12640 (Area blow-up)
Very unusual crater, which has signs of an atomic explosion

Look towards middle left. Very unusual object within crater. (NASA photo Goclenius crater centre bottom).

*Unknown crater PLATE 88 (APOLLO 8).
Notice black line to centre of picture, a bridge perhaps?
And unusual cross middle to right. Mining?*

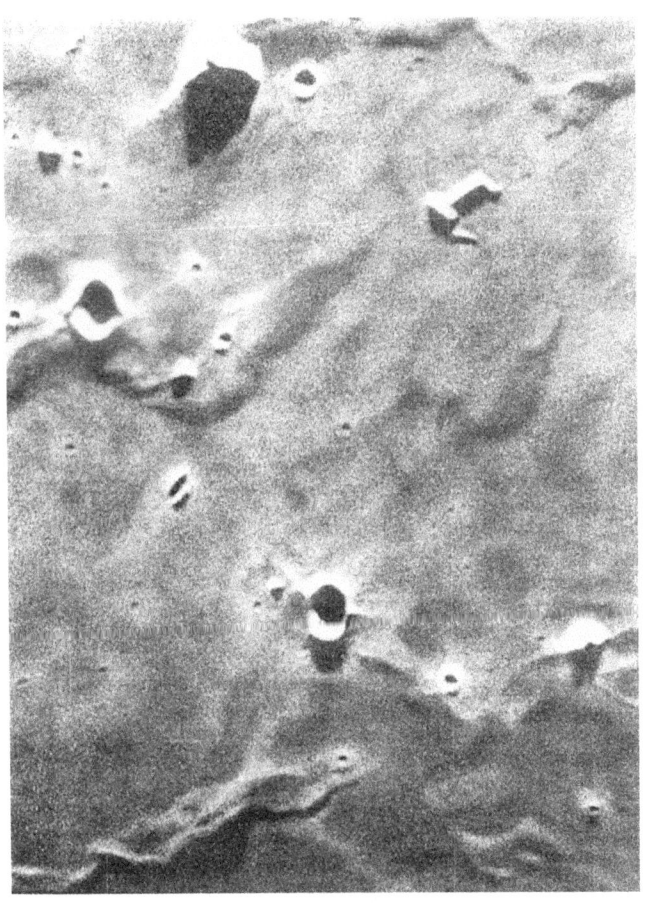

Strange shape top right (Photograph by Russian orbiter). Investigated by myself and Neil Hunter 1990's.

Notice on both pictures the letter S.

Four frames taken from Apollo 16, as Duke and Young briefly talk about footprints.

The Short Movie Script

This is a script my wife and myself put together back in the 90's for a short movie based on an incident on the moon. The script is actually using a lot of true facts and is based on what might have happened already, what could happen or what is happening, anyway please read and enjoy,

Simon Lewis

The Two Faced Moon

An Original Screenplay by
Carol and Simon Lewis

A spaceship orbits the moon. The spacecraft has the words LUNA 13 on its side. It is piloted by two Astronauts, James Lewis and Jack King, both very experienced in Luna exploration. The year is 1976. 24th June. 2100 hours.

LEWIS
Mission Control, Mission Control, do you copy?
MISSION CONTROL
Receiving you Lewis.
LEWIS
We are approaching the dark side. 2 minutes to radio silence. Advise over.
MISSION CONTROL
All systems are looking good. You are OK for descent.
LEWIS
Receiving you loud and clear.

The small spacecship now fires small thrusters which allow it to position for descent to the moon's surface. They plan to land in a crater known as Tsiokovshi, 150 miles in diameter, with steep walls and a massive central mountain area. This is considered to be a very dangerous mission.

LEWIS
10 seconds to radio silence.

MISSION CONTROL
Good luck Lewis, King. Hope the landing is a good one.

KING

Speak to you in 5 days. Changing over to auto-pilot.

The spaceship breaks contact and starts to descend. A central rocket fires as the craft makes its way towards the surface. The pilots continue to operate and check controls as it runs on auto-pilot.

LEWIS

When we're down there King, I'll be a lot happier. This is one hell of a ride.

KING

Yeah, I'll be glad when we're there too, this ship certainly is a bone shaker.

LEWIS

30 minutes to landing and counting.

KING

Main thrusters OK, pressure OK, computer readings OK.

The spaceship is now very close to the surface. The craters are now huge.

LEWIS

15 minutes to landing. What's your reading on pressure levels?

KING

Slight drop in pressure. Main thrusters OK. Positioning thrusters hanging in there. Computer reading are showing some unusual disturbances out there.

LEWIS
Probably nothing to worry about. Do you remember in practice, we had worse than this.

KING
Yeah, guess your right.

The ship suddenly loses control. There's a small explosion on the outside. Oxygen starts to escape into space.

LEWIS
My God King, what was that?

KING
I don't know, but turn off the auto-pilot. We're losing pressure. Thrusters are not looking good. Electrical malfunctions.

LEWIS
OK, what about oxygen?

KING
Bad news, we need to get to the surface.

Both pilots press switches and look at a computer screen. The spaceship shakes violently.

LEWIS
We have 5 minutes to impact. Thrusters are almost gone. And that damn Tsiokovshi crater is coming at us like a raging bull.

KING
What about side thrusters?

LEWIS
No good, no good.

KING
Well there must be something. We're not going down without a fight.

LEWIS
Suggest something soon. Oh, God.

The craft suddenly stops shaking. There is a strange silence.

LEWIS
King?

KING
What's happened? I didn't do anything.

LEWIS
Neither did I, but look out of the window.

King looks towards a small window and sees a huge cigar shaped object. Still and silent.

KING
Am I seeing things or is that what I think it is?

LEWIS
Looks as though we're not the only ones who came here. And that's probably an understatement.

KING
Well understatement or not, something has just saved our lives and is not going anywhere.

LEWIS
Do you think it's ours?

KING
Well we've been briefed on this sort of thing. Never dreamed it would ever happen.

LEWIS
If it is ours, we're flying a dustbin in comparison to that beauty. I wonder if we should try to make contact.

King and Lewis continue to peer out of the window. The ship is so big, their spaceship is tiny in comparison. The ship has many windows and huge round openings. In them can be seen smaller disc shaped ships.

Lewis and King try to make contact.

LEWIS
I don't really know how to start. What should I say?

KING
Thank you sounds like a good idea.

LEWIS
Guess your right.

Lewis fiddles with a few switches and speaks.

LEWIS
Hello, hello. Can anybody hear me? Do you understand English? We'd like to thank you for saving our lives.

KING
No response Lewis. But there is movement. The radar shows something coming towards us. It's much smaller and disc-shaped.

LEWIS
Looks like contact has been made.

Both men look at each other. Around them lights flash and flicker.

LEWIS
Funny isn't it. Here we are, so far from Earth, no contact with them. We've just avoided hitting the moon and now we face, well something.

Suddenly there is a tapping noise. Both men look to the spaceship entrance.

KING
This is it then. It looks as though our saviours are knocking at the door.

LEWIS
Lets hope they are friendly.

The door makes a loud hissing noise. Both men look on as the door opens. As soon as the door opens they have gravity. Lewis steps forward. A small being, grey in appearance, 4ft tall with large black eyes, small nose, no hair walks towards them both.

LEWIS
Welcome. We are grateful for your assistance. Do you speak English? We mean no harm, we come in peace.

The small being is joined by another from behind. Their lips are small but seem to move. The first being holds out it's hand which has 3 fingers and what looks like a thumb. Lewis moves closer. They both join hands.

LEWIS
I think they're friendly.

The first being smiles. The second being points to the hatch. The men follow. The inside of the ship is small with no visible control. Just two small seats and a rounded ceiling of silver. Both men are shown an area to sit, rectangular seats appear from the wall of the ship. They sit down. The hatch closes. Movement of the ship is felt. It moves towards a large opening in the cigar shaped ship. Their spaceship is also moved somehow with a kind of energy field.

LEWIS
I wonder where they are taking us.

The ship suddenly jerks. A small click noise is heard. And the door to the craft opens. The two beings stand and one of them points to the door.

LEWIS
I think it's time to move. Do you notice, there's still gravity.

KING
Yeah I wonder how they got round that problem. Certainly beats floating around our ship.

Both men walk into a huge room where standing is a humanoid. Both are quite surprised.

LEWIS
I don't believe it, are you one of us?

JARKAN
Greetings. Welcome to our ship. Please do not be afraid, I can assure you, you are safe here.

KING

Do you have a name, and what about the two little chaps.

JARKAN

My name is Jarkan, and my small friends are my helpers. Or in your understanding, robots. They do not have names but numbers, which you will understand as Helper 8 and Helper 9.

The two small beings stand silently as Jarkan speaks.

LEWIS

I have to ask you just one question, are you from Earth?

JARKAN

No. I am from here.

KING

This is your world then?

JARKAN

Please, all your questions will be answered later. You are our guests and maybe you would like to freshen up before we speak again.

Jarkan turns and walks away. The two small beings point to a long passage. Lights run all the way down. The men follow the beings.

> **LEWIS**
> This ship is amazing. Look at the space. You'd get 25 of our ships in here with room to spare.

The men are shown to a small room. Suitable clothing, water and food are there. The huge cigar shaped object is now moving towards a heavily cratered area of the moon. A triangular crater comes into view. Masses of light come from it as the ship descends. Around the unusual opening machines are at work. Inside there is a giant underground city. Lights shine from unusual structures and other objects fly within. The ship makes it's way to a large structure where it connects and comes to a halt. In the meantime, Lewis and King are freshened and clothed in unusual outfits and are led from the ship into the main building. They are then taken to a small room. Inside Jarkan greets them.

> **JARKAN**
> Please be seated. I'm sure you need an explanation. Please be patient and I'll answer all your questions.

The two small grey beings are stood to the side. Jarkan stands with a huge window behind him which has a view of the underground city.

> **JARKAN**
> Your ship was badly damaged but is being repaired.

KING
I guess we should be grateful that you turned up or we might not be alive now.

JARKAN
Yes gentlemen, but I'm afraid we were responsible for the damage to your ship.

LEWIS
What do you mean exactly?

JARKAN
You were going to land in an area where we are carrying out a lot of mining work and exploration. It would not have been good for you to land without notice.

KING
I'm confused, you damaged our ship? And you're mining here? And all this technology?

JARKAN
We have been watching your progress as a race for a long time. You've advanced very quickly, especially in the last 100 years.

LEWIS
You mean you have been here that long?

JARKAN
Over two thousand years in fact.

KING
But why are you here?

JARKAN
I would like to ask you the same question? You have become very worrying as a race. War and power seem to be your main problem.

KING
But we are not all power hungry and killing each other.

JARKAN
Yes I understand that. But the people in power abuse and use the rest.

LEWIS
If you have been here so long, and watched us, why have you not helped us to evolve better?

JARKAN
We have tried on many occasions. Our ancestors landed on earth and communicated with many different cultures. We were received as what you call Gods. Which sadly followed by war in most cases, between themselves.

KING
What you're telling us, certainly fits in with many baffling stories and mysteries of our past, some never solved.

JARKAN

Yes, we probably had a lot to do with many situations that have occurred on your planet, but not just by us alone. There are about 25 thousand different beings spread among the stars and many have been to your world long before we arrived.

LEWIS

What about God and religions. Do you have any answers?

JARKAN

Your God as you call it is the ultimate energy force and does create worlds and life. But as far as we know it is so far in advance than ourselves, we don't understand how it all works. You see the universe is far more complicated than we and you can ever imagine. Life as we understand it is a body and a soul. But passed that point there is an energy that is a life. But far more advanced than even we can imagine. Please gentlemen, we can talk about these things another time. Perhaps we should show you our world. Please look towards the window.

The glass window suddenly turns to a picture and different scenes start to appear.

JARKAN

This is our city. The structures are

mainly made of glass and elements only known to us. The larger ship you can see moving out is our starship. Capable of travelling unimaginable distances. Powered by what we call clean energy. The small glass buildings are where we keep our plants, trees and food. All plant based. We have built glass structures before on the surface but due to the unpredictable meteor showers it created too many problems. The disc shaped craft is well known to you as a flying saucer and is used for short distances of travel. As you've already seen, it holds two helpers. We ourselves use the much larger ships. The helpers do most of our work.

Pictures are shown of humanoids and helpers, farming, mining and relaxing.

JARKAN
So gentlemen, this is our world. And has been for a long time. But we did live as you did a long time ago.

The screen disappears, the window is now restored.

LEWIS
Can you imagine if they all saw this on earth.

JARKAN
They already know. But only in your

higher places. We have found them to be very difficult to deal with.

LEWIS
What do you mean?

JARKAN
We have had allsorts of accidents. The first major one you know as Roswell. 1947 was your year.

KING
But that was supposed to be a weather balloon.

JARKAN
Two of our helpers ran into trouble and had to crash land. Your military soon captured them. Our cover was broken. Negotiations took place. Technology changed hands. You advanced very quickly with our help. How do you think you got here today?

LEWIS
You mean you helped design our ship?

JARKAN
In a way, most of the design goes back thousands of years to our early spacecraft.

KING
Did they know on earth about us coming here.

JARKAN
You ask so many questions.

KING
But did they know?

Jarkan turns and looks at the two helpers. One steps forward and speaks. Both men are suprised.

HELPER 8
The answer is yes. You two were chosen for your expertise and knowledge in Luna exploration. Unfortunately, your landing coordinates where planned by people unaware of your governments treaty with us.

KING
What treaty?

HELPER 8
You were unaware of the purpose for your mission for security reasons. The real reason for your mission is for us to impart our knowledge to your peoples. You are trained in space flight. You are from Earth. You are the next step forward to bringing peace to your world. With our help, your world could become a better place.

LEWIS
All the while you are saying this, but you have already made it clear that we

are power mad and always at war.

HELPER 8
You don't have a choice. Your world is dying. Your energies are nearly depleted. Your power base is collapsing. War could be your end. But not before world famine globally effects you all.

LEWIS
Why save us all if we have made such a mess?
We surely cannot be trusted.

HELPER 8
If your world goes, our world goes. They are but one and one effects the other.

KING
You mean finally we have nearly done the unthinkable and reached our end?

Jarkan steps forward and Helper 8 moves back.

JARKAN
Yes your planet is badly damaged. Not just with the environment but war which seems to be everywhere. We have tried to help recently but communications can be difficult at times with your governments. When your ship is repaired you will leave here. But when reaching the other side of the moon

> your mission control will contact you.
> You will transmit a message and with
> that will be a film telling your people
> all about us. It will reach all screens
> across your world. Your people will see
> that you have to change for the good of
> all our races.

Lewis and King look at each other for a long moment and the scene ends.

THE FINAL SCENE

The men are seen next in their ship, moving away from the triangular base back into the moons orbit. The ship reaches radio contact.

> **LEWIS**
> Mission control, do you read?
>
> **MISSION CONTROL**
> Yes we hear you loud and clear.
>
> **LEWIS**
> Glad to hear your voices at last. We've had an amazing time but I guess you knew that was coming.
>
> **MISSION CONTROL**
> Did you get the information?
>
> **LEWIS**
> Yes sir. Ready to transmit.

Lewis looks at King. They smile. King presses

a button. There is a huge explosion. The spaceship explodes into thousands of pieces. We return to mission control. Or what they thought was Mission Control. The signal had been intercepted on a secret channel.

Back at the moon city, Helper 8 and Helper 9 are sitting in a room. Helper 8 touches a panel in front of him.

HELPER 8
Mission control to Lewis do your read?

HELPER 9
Mission accomplished.

www.ingramcontent.com/pod-product-compliance
Ingram Content Group UK Ltd.
Pitfield, Milton Keynes, MK11 3LW, UK
UKHW021043090125
453260UK00011B/393